Issues in Commuting and Pilot Fatigue: Interim Report

Committee on the Effects of Commuting on Pilot Fatigue

Board on Human-Systems Integration

Division of Behavioral and Social Sciences and Education

Transportation Research Board

NATIONAL RESEARCH COUNCIL
OF THE NATIONAL ACADEMIES

THE NATIONAL ACADEMIES PRESS
Washington, D.C.
www.nap.edu

THE NATIONAL ACADEMIES PRESS 500 Fifth Street, N.W. Washington, DC 20001

This study was supported by Grant No. DTFAWA-10-C-00115 between the National Academy of Sciences and the Federal Aviation Administration. Any opinions, findings, conclusions, or recommendations expressed in this publication are those of the author(s) and do not necessarily reflect the views of the organizations or agencies that provided support for the project.

International Standard Book Number -13: 978-0-309-18712-1
International Standard Book Number -10: 0-309-18712-5

Additional copies of this report are available from National Academies Press, 500 Fifth Street, N.W., Lockbox 285, Washington, DC 20055; (800) 624-6242 or (202) 334-3313 (in the Washington metropolitan area); Internet, http://www.nap.edu.

Printed in the United States of America

Suggested citation: National Research Council (2011). *Issues in Commuting and Pilot Fatigue: Interim Report.* Committee on Commuting and Pilot Fatigue, Board on Human-Systems Integration, Division of Behavioral and Social Sciences and Education. Washington, DC: The National Academies Press.

THE NATIONAL ACADEMIES

Advisers to the Nation on Science, Engineering, and Medicine

The **National Academy of Sciences** is a private, nonprofit, self-perpetuating society of distinguished scholars engaged in scientific and engineering research, dedicated to the furtherance of science and technology and to their use for the general welfare. Upon the authority of the charter granted to it by the Congress in 1863, the Academy has a mandate that requires it to advise the federal government on scientific and technical matters. Dr. Ralph J. Cicerone is president of the National Academy of Sciences.

The **National Academy of Engineering** was established in 1964, under the charter of the National Academy of Sciences, as a parallel organization of outstanding engineers. It is autonomous in its administration and in the selection of its members, sharing with the National Academy of Sciences the responsibility for advising the federal government. The National Academy of Engineering also sponsors engineering programs aimed at meeting national needs, encourages education and research, and recognizes the superior achievements of engineers. Dr. Charles M. Vest is president of the National Academy of Engineering.

The **Institute of Medicine** was established in 1970 by the National Academy of Sciences to secure the services of eminent members of appropriate professions in the examination of policy matters pertaining to the health of the public. The Institute acts under the responsibility given to the National Academy of Sciences by its congressional charter to be an adviser to the federal government and, upon its own initiative, to identify issues of medical care, research, and education. Dr. Harvey V. Fineberg is president of the Institute of Medicine.

The **National Research Council** was organized by the National Academy of Sciences in 1916 to associate the broad community of science and technology with the Academy's purposes of furthering knowledge and advising the federal government. Functioning in accordance with general policies determined by the Academy, the Council has become the principal operating agency of both the National Academy of Sciences and the National Academy of Engineering in providing services to the government, the public, and the scientific and engineering communities. The Council is administered jointly by both Academies and the Institute of Medicine. Dr. Ralph J. Cicerone and Dr. Charles M. Vest are chair and vice chair, respectively, of the National Research Council.

www.national-academies.org

COMMITTEE ON THE EFFECTS OF COMMUTING ON PILOT FATIGUE

Clinton V. Oster, Jr. (*Chair*), School of Public and Environmental Affairs, Indiana University
Benjamin A. Berman, Senior Research Associate, Ames Research Center, U.S. National Aeronautics and Space Administration
J. Lynn Caldwell, Senior Research Psychologist, Wright-Patterson Air Force Base, Ohio
David F. Dinges, Department of Psychiatry, University of Pennsylvania School of Medicine
R. Curtis Graeber, The Graeber Group, Kirkland, Washington
John K. Lauber, Independent Consultant, Vaughn, Washington
David E. Meyer, Department of Psychology, University of Michigan
Matthew Rizzo, Department of Neurology, Mechanical and Industrial Engineering, and the Public Policy Center, University of Iowa
David J. Schroeder, Independent Consultant
J. Frank Yates, Department of Psychology, University of Michigan

Mary Ellen O'Connell, *Project Director*
Toby Warden, *Study Director*
Eric Chen, *Senior Project Assistant*
Stephen Godwin, *Liaison, Studies and Special Programs,* Transportation Research Board

BOARD ON HUMAN-SYSTEMS INTEGRATION

William S. Marras (*Chair*), Integrated Systems Engineering Department, Ohio State University
Pascale Carayon, Department of Industrial and Systems Engineering, Center for Quality and Productivity Improvement, University of Wisconsin–Madison
Don Chaffin, Industrial and Operations Engineering and Biomedical Engineering, University of Michigan (*Emeritus*)
Nancy J. Cooke, Cognitive Science and Engineering, Arizona State University
Mary (Missy) Cummings, Aeronautics and Astronautics and Engineering Systems Division, Massachusetts Institute of Technology
Sara J. Czaja, Department of Psychiatry and Behavioral Sciences, Center on Aging, University of Miami Miller School of Medicine
Andrew S. Imada, A.S. Imada and Associates, Carmichael, California
Waldemar Karwowski, Department of Industrial Engineering and Management Systems, University of Central Florida
David Rempel, Department of Medicine, University of California, San Francisco
Matthew Rizzo, Department of Neurology, Mechanical and Industrial Engineering, and the Public Policy Center, University of Iowa
Thomas B. Sheridan, Departments of Mechanical Engineering and of Aeronautics-Astronautics, Massachusetts Institute of Technology (*Emeritus*)
David H. Wegman, Department of Work Environment, University of Massachusetts, Lowell (*Emeritus*)
Howard M. Weiss, Department of Psychological Sciences, Purdue University

Barbara A. Wanchisen, *Director*
Mary Ellen O'Connell, *Deputy Director*
Christie R. Jones, *Program Associate*

ACKNOWLEDGMENTS

The report has been reviewed in draft form by individuals chosen for their diverse perspectives and technical expertise, in accordance with procedures approved by the National Research Council's Report Review Committee. The purpose of this independent review is to provide candid and critical comments that will assist the institution in making its published report as sound as possible and to ensure that the report meets institutional standards for objectivity, evidence, and responsiveness to the study charge. The review comments and draft manuscript remain confidential to protect the integrity of the deliberative process. We wish to thank the following individuals for their review of this report: Evan Byrne, Human Factors Group, National Transportation Safety Board; James C. Miller, Human Factors Consultant, San Antonio, Texas; Joseph P. Ornato, Department of Emergency Medicine, Virginia Commonwealth University, Richmond, Virginia; and Nita Lewis Shattuck, Human Systems Integration Program, Operations Research Department, Naval Postgraduate School, Monterey, California.

Although the reviewers listed above have provided many constructive comments and suggestions, they were not asked to endorse the conclusions or recommendations nor did they see the final draft of the report before its release. The review of this report was overseen by Deborah A. Boehm-Davis, Psychology Department, George Mason University as review coordinator and Floyd E. Bloom, Molecular and Integrative Neuroscience Department, The Scripps Research Institute as review monitor. Appointed by the National Research Council, they were responsible for making certain that an independent examination of this report was carried out in accordance with institutional procedures and that all review comments were carefully considered. Responsibility for the final content of this report rests entirely with the authoring committee and the institution.

The committee would also like to thank the individuals who provided presentations to the committee during this initial phase of the study; see Appendix B for the open agendas of the first two meetings.

Contents

EXECUTIVE SUMMARY

The potential for fatigue to negatively affect human performance is well established. Concern about this potential in the aviation context extends back decades, with both airlines and pilots agreeing that fatigue is a safety concern. A more recent consideration is whether and how pilot commuting, conducted in a pilot's off-duty time, may affect fatigue.

The Airline Safety and Federal Aviation Administration Extension Act of 2010 (P.L. 111-216) directed the Federal Aviation Administration (FAA) to contract with the National Academy of Sciences to conduct a study on the effects of commuting on pilot fatigue. The study is intended to inform the development of commuting-related aspects of FAA regulations also specified in the act.

The committee was asked to review available information related to the prevalence and characteristics of pilot commuting; sleep, fatigue, and circadian rhythms; airline and regulatory oversight policies; and pilot and airline practices. Based on this review, the committee will define and discuss several related topics:

- commuting in the context of pilot alertness and fatigue;
- the relationships between the available science on alertness, fatigue, sleep, and circadian rhythms, cognitive and physiological performance, and aviation safety;
- the policy, economic, and regulatory issues that affect pilot commuting;
- the commuting policies of commercial air carriers and, to the extent possible, practices that are supported by the available research; and
- potential next steps, including, to the extent possible, recommendations for regulatory or administrative actions, or further research, by the FAA.

This interim report summarizes the committee's review to date of the available information. The final report will present a final review, along with the committee's conclusions and recommendations based on the information available during its deliberations.

Fatigue has multiple interactive sources. The primary ones that may be relevant to pilot commutes include: duration of time awake prior to work, duration of time slept prior to work, quality or restfulness of sleep (i.e., sleep continuity) prior to work, and the biological time (i.e., circadian phase) at which commuting occurs relative to start of work. The duration of time at work (i.e., time on task) is a regulated factor for fatigue mitigation, but it is not regulated relative to commute time since the latter is off-duty time. In the aviation industry, commutes that involve travel across multiple time zones have the potential to exacerbate fatigue associated with commuting, as can chronic restriction of sleep for multiple days prior to commuting.

Commuting by pilots is a common practice that is characterized by tremendous variability. However, comprehensive information on the prevalence or characteristics of commuters is currently unavailable. Although extensive scientific research has been conducted on alertness, fatigue, sleep and circadian rhythms, cognitive and physiological performance, and safety—including research specific to the aviation industry—there is a paucity of information on the nature of commutes or how commuting affects factors that connect sleep and performance.

The committee's charge does not include a systematic survey of either pilots or airlines, and the specified time and available resources also preclude such a survey. Instead, the committee has requested relevant information from pilot and airline associations, consumer groups, and individual airlines. That information will be considered, along with further analysis of the relevant research and its implications given what is known about the aviation industry and commuting, in the committee's final report. To date, the extent and circumstances under which commuting contributes to fatigue are unclear.

1
INTRODUCTION

Nearly everyone experiences fatigue, but some professions, such as aviation, medicine and the military, demand alert, precise, rapid, and well-informed decision making and communication with little margin for errors. Recognizing this, the National Transportation Safety Board (NTSB) added "Reduce Accidents and Incidents Caused by Human Fatigue in the Aviation Industry" to its list of most wanted aviation safety improvements two decades ago. Specifically, the NTSB called for research, education, and revisions to regulations related to work and duty hours. Regulatory change has received the least attention, with no changes to relevant regulations since 1985 despite a significantly expanded research base on sleep, fatigue, and circadian rhythms.[1]

Concern about the potential contribution to fatigue from time spent commuting to a duty station was elevated following a fatal Colgan Air crash in Buffalo, New York, February 12, 2009. The crash, and the first officer's cross-country commute, received substantial media attention. The NTSB determined that the probable cause of the accident was "the captain's inappropriate response" to a low speed condition (National Transportation Safety Board, 2010, p. 155). The NTSB report identified multiple contributing factors related to flight crew and corporate factors, but did not list fatigue or commuting as a contributing factor or cause in the accident. Instead, the Board concluded that "the pilots' performance was likely impaired because of fatigue, but the extent of their impairment and the degree to which it contributed to the performance deficiencies that occurred during the flight cannot be conclusively determined" (National Transportation Safety Board, 2010, p. 108).

Against this backdrop, in September 2010 Congress directed the Federal Aviation Administration (FAA) to revise its regulations related to work and duty hours to reflect current research (P.L. 111-216). The law also directed the FAA to contract with the National Academy of Sciences (NAS) to conduct a study of the effects of pilot commuting on fatigue. The NAS was directed to review information in seven specified areas. Based on that review, the NAS was charged to discuss relevant issues with the goal of identifying potential next steps, including possible recommendations related to regulatory or administrative actions or further research that can be taken by the FAA: see Box 1. The FAA issued a Notice of Proposed Rulemaking (NPRM) on September 14, 2010, inviting public comment that would be considered in issuing final regulations. The NAS study is designed to inform the component of these final regulations relevant to pilot commuting.

The NAS established the Committee on the Effects of Commuting on Pilot Fatigue (see Appendix A) to conduct this study. This interim report highlights key findings from the scientific literature on fatigue in relation to time awake, time asleep, and time of day; identifies issues the committee will need to consider; specifies the information available to the committee to date; and presents the committee's plans for collecting additional information. The committee's final report, expected to be issued in summer 2011, will present its conclusions and

---

[1]A Notice of Proposed Rulemaking (NPRM) with revised regulations on this topic was promulgated in 1995, but it was withdrawn in 2009 with the acknowledgment that changes since 1995 in both the world of commercial aviation and the scientific understanding of fatigue had rendered it out of date. A new rulemaking activity was started that resulted in a new NPRM issued in 2010. The committee discusses this FAA rulemaking activities related to flight crew fatigue in a subsequent part of this report.

recommendations based on the information available during the course of its deliberations. It should be noted that this interim report is being provided according to tasking requirements, and that nothing in this interim report should be construed as findings, conclusions, or recommendations from this committee regarding pilot commuting. Rather, it is intended to provide a snapshot of the committee's activities to date, and to present in broad terms a "road map" of how the committee intends to address the issues in its final report.

BOX 1
Information to Be Reviewed and Study Objectives

The study is to review available information on:

- the prevalence of pilots commuting in the commercial air carrier industry, including the number and percentage of pilots who commute greater than two hours each way to work;
- characteristics of commuting by pilots, including distances traveled, time zones crossed, time spent, and methods used;
- the impact of commuting on pilot fatigue, sleep, and circadian rhythms;
- commuting policies of commercial air carriers (including passenger and all-cargo air carriers), including pilot check-in requirements and sick leave and fatigue policies;
- postconference materials from the Federal Aviation Administration's June 2008 symposium titled "Aviation Fatigue Management Symposium: Partnerships for Solutions";
- Federal Aviation Administration and international policies and guidance regarding commuting; and
- to the extent possible, airline and pilot commuting practices.

Based on this review, the committee will:

- define "commuting" in the context of pilot alertness and fatigue;
- discuss the relationship between the available science on alertness, fatigue, sleep and circadian rhythms, cognitive and physiological performance, and safety;
- discuss the policy, economic, and regulatory issues that affect pilot commuting;
- discuss the commuting policies of commercial air carriers and to the extent possible, identify practices that are supported by the available research; and
- outline potential next steps, including to the extent possible, recommendations for regulatory or administrative actions, or further research, by the Federal Aviation Administration (FAA).

2
APPROACH TO INFORMATION COLLECTION

There is extensive research—including research specific to the aviation industry—on alertness, fatigue, sleep and circadian rhythms; cognitive and physiological performance; and safety. However, there is very little information specifically on pilot commuting, including commuting practices or airline policies and practices related to commuting. To help address this gap, the committee issued a call for input that was sent to pilot and airline associations and passenger groups and was posted on the project website: see Box 2. The people and groups involved were invited to respond to a series of questions specific to the types of information the committee was asked to review: see Box 3.

BOX 2

Organizations Contacted for Input

Pilot Associations and Unions

- Air Line Pilots Association
- Coalition of Airline Pilots Associations
- Allied Pilots Association (American Airlines pilots)
- Independent Pilots Association (UPS pilots)
- Southwest Airlines Pilots Association
- Teamsters Local 1224 (Horizon Air, Southern Air, ABX Air, Atlas Air, Polar Air Cargo, Atlas Worldwide, Kalitta Air, Cape Air, Miami Air, Gulfstream Air, Omni Air and USA 3000 pilots)
- US Airline Pilots Association (US Airways pilots)
- International Federation of Airline Pilots Association

Airline Associations

- Air Transport Association
- Cargo Airline Association
- Federal Express
- National Air Carrier Association
- National Business Aviation Association
- National Air Transport Association
- Regional Air Cargo Carriers Association
- Regional Airline Association
- UPS Airlines

Groups That Represent Passenger Interests

- Air Travelers Association
- Flight Safety Foundation

In addition, the request for input was posted on the project website, http://www7.nationalacademies.org/bbcss/public_form_invitation.doc.

BOX 3
Topics Posed in Call for Public Input

Interested organizations or individuals were invited to provide comments on their perspective in the following areas, as relevant to their work and experience:

(A) the prevalence of pilots commuting in the commercial air carrier industry, including the number and percentage of pilots who commute greater than two hours each way to work;

(B) the characteristics of commuting by pilots, including distances traveled, time zones crossed, time spent, and methods used;

(C) the impact of commuting on pilot fatigue;

(D) whether, and if so how, the commuting policies and/or practices of commercial air carriers (including passenger and all-cargo air carriers), including pilot check-in requirements and sick leave and fatigue policies, ensure that pilots are fit to fly and maximize public safety;

(E) whether, and if so how, pilot commuting practices ensure that they are fit to fly and maximize public safety;

(F) how "commuting" should be defined in the context of the commercial air carrier industry; and

(G) how FAA regulations *related to commuting* could or should be amended to ensure that pilots arrive for duty fit to fly and to maximize public safety.

This report reflects comments received to date; additional information received during the course of the committee's work will be included in the final report. The committee also invited interested organizations and individuals to present their responses both in writing and in person at meetings held in November and December 2010 (see Appendix B for the public meeting agendas).

The committee also is assessing information from the following sources:

- a review of NTSB reports on aviation accidents to identify available information related to the contribution of commuting to flight crew fatigue;
- a review of confidential reports mentioning commuting and/or fatigue submitted to the Aviation Safety Reporting System (ASRS), a voluntary pilot reporting system funded by the FAA and hosted by NASA;
- a review of the comments related to commuting or fitness for duty submitted in response to the NPRM;
- a review of available information on relevant airline policies and practices in the international arena;
- analysis of data requested from airlines on pilot residence (approximated by zip code) and duty location (domicile or base) to enable an approximation of commuting distance and time; and
- a review of the relevant scientific literature.

3
COMMUTING IN THE AVIATION CONTEXT

For most people, commuting is a simple concept that represents the daily time spent traveling—almost always by ground transportation—from their homes to their workplaces. For pilots, the meaning of "commuting" is often more complex.

First, it is not uncommon for pilots to travel by air to and from their flight assignment. Commuting enables pilots to live a considerable distance from the airport at which they are based and travel to work in a relatively short time. For example, a two-hour commute via air would enable a pilot to live a considerably further distance than if the commute were by land. Second, commuting is not typically a daily occurrence, as pilot duty assignments often extend over several days and keep pilots away from home for multiple days at a time. As a result, a pilot's commute to work may be undertaken as infrequently as once or twice per month—or more frequently, depending on the pilot's flying schedule and commuting arrangements. Third, pilots sometimes travel to arrive nearby their domicile (the location of the base from which they fly) for a period before they are scheduled to fly, for logistical reasons, to have a rest opportunity, or for both reasons. Whether commuting time should be considered to start when pilots leave their homes or the place where they last slept is still under consideration by the committee. The key issue is whether a pilot begins the subsequent duty rested and fit to fly.

In addition to the lack of data on the prevalence of pilot commuting noted above, there are few data on specific methods or other characteristics of pilot commuting. Furthermore, all commutes, even commutes involving the same amount of time, may not have the same potential to influence fatigue. Some commutes may not be cognitively or physically demanding (e.g., seated as a passenger on a train, bus, or plane), even to the point of permitting sleep to be obtained, while other commutes may entail more physical (e.g., standing) or cognitive (e.g., driving) demands.

For the purposes of this interim report, the committee is considering the following working operational definitions and issues in defining pilot commuting. These definitions and issues will be evaluated as the committee acquires more information, and they may be refined or changed in the final report.

The committee considers a pilot's "domicile" to be the airport where a pilot begins and ends a duty period. This is distinguished from "hub," which is a focus for the routing of aircraft and passengers. A pilot's domicile may be at one of the airline's hubs, but it may also be at an airport that does not serve as a hub for that airline.

The committee considers a pilot's "home" to be the pilot's residence: it is important to note that it is not necessarily the place where the pilot had the most recent opportunity for his or her customary sleep period. For example, the pilot may have access to a hotel room, apartment, or other sleep accommodation near his or her domicile.

The committee considers pilot "commuting" to be the period of time and the activity required of pilots from leaving "home" to arriving at the domicile (airport—in the crew room, dispatch room, or designated location at the airport) and from leaving the domicile back to "home." Distinguishing a commute from a noncommute solely on the basis of the duration or distance of commute, or modality of commuting, will require closer examination and may be overly simplistic. Such an approach may be desired for operational applications, but the scientific foundations for establishing such a taxonomy relative to fatigue are not yet nearly as well founded as the scientific literature on fatigue in relation to time wake, time asleep, and time of day.

In its charge, the committee was asked to distinguish commutes of greater than 2 hours from other commutes. This dividing line, though potentially arbitrary,[2] will be examined if the committee can obtain relevant data

4
PREVALENCE OF COMMUTING

Although it should be the cornerstone of the committee's review, the committee has yet to uncover any systematic or comprehensive data[3] on either the frequency of pilot commuting, the length of the commute, or the characteristics of commuting by pilots. The comments the committee has received to date from both pilots and airlines supports the view that pilot commuting is an integral and necessary aspect of the commercial aviation industry in the United States. A case could be made that the committee should ideally acquire systematic data to quantify the prevalence and characteristics of pilots and their commutes. However, developing, testing, implementing, and analyzing a pilot survey to acquire such data would require an

---

[2]The FAA's recently published NPRM incorporates the suggestion that a "local area" be defined as an area within a two-hour travel period, regardless of mode of transportation.

[3]The only published information appears to be data included in the NTSB report following the Colgan Air crash, which reported that 68% of the Colgan pilots based at Newark were commuting, with the commutes being various distances (NTSB, 2010; pp. 47-48).

extended timeframe that well exceeds the time and resources available to the committee. Instead, the committee is relying on data it can obtain from the aviation sources (i.e., NTSB, ASRS, airlines, pilot associations) mentioned above. This effort is still in progress.

In addition to the general call for information, outlined in Box 3, the committee also requested information from individual Part 121 airlines, using a list of airlines provided by the FAA. Part 121 applies to most passenger and cargo airlines that fly transport-category aircraft with ten or more seats. Specifically, airlines were asked to provide data on pilots by domicile and home zip code: such data would enable the committee to obtain an individual-level approximation of commuting. The committee hopes to obtain data from multiple airlines, but very little information was available by the time of this interim report. Input to date, however, suggests that commuting more than 2 hours is not uncommon among pilots.

5
AVIATION INDUSTRY CHARACTERISTICS THAT IMPACT COMMUTING

Characteristics of the aviation industry that influence pilot commuting include airline crew scheduling practices; airline route network and crew basing practices; and airline competitive and passenger demand factors that can cause pilot staffing requirements to change over time. These characteristics then interact with pilots' preferences related to commuting and may influence their decisions about where to maintain residency. Certain airline policies and practices can facilitate or impede a pilot's ability to commute, but at this point in the study, it is unclear how such policies potentially affect fatigue that may result from the commuting activity.

Airline Crew Scheduling: A Pilot's Work Pattern

At most airlines, labor agreements between pilots and airlines establish specific policies and practices regarding flight crew scheduling (within requirements for flight and duty time as defined in Federal Aviation Regulations). Virtually all of these airlines rely on a bidding process to award monthly schedules (sometimes called lines or blocks) to pilots; selection advantages are given to pilots on the basis of seniority. Typically, a monthly schedule consists of multiple assignments of trips (sometimes called pairings), each of which may consist of several flights over a period lasting 1, 2, or up to more than 6 days. Each of these trips begins and ends at the pilot's domicile (there also may be one or more overnights elsewhere) and thus comprises the basic duty assignment to and from which the pilot commutes.

While a "9 to 5" worker may commute on a daily basis, airline pilots may commute much less frequently while also remaining away from home for multiple days at a time on each of these trips. By federal regulations, airline pilots are limited to fly no more than 1,000 hours per year, or an average of about 83 hours per month. On the basis of this monthly limit, the number of flight hours per trip will determine the number of trips—and thus, potentially, the number of commutes—during the month. For example, if a pilot's trips involve 20 hours of flying over 4 days, the pilot will do about four of these trips per month. There will be one or more days off between each trip. Other factors, such as flight cancellations and delays, have the potential to influence the length of a trip and the time off between assignments and, therefore, subsequent trips and commutes.

Using the seniority-based bidding process, pilots select the desired trips and days worked given their individual preferences, including the nature of their commutes, if any. For example, a pilot who commutes a long distance by air to the domicile may bid for the monthly line of four, 4-day trips, specifically, trips beginning at the domicile late on the first day (allowing for the inbound commute) and ending back at the domicile relatively early on the fourth day (allowing for the homebound commute). This pilot will make four commutes during the month. In contrast, a pilot who lives near the domicile (e.g., driving 45 minutes to the airport) may bid for ten 1-day trips, each of which starts early in the morning and returns to the domicile later that day. This pilot will make ten commutes during the month. Note that the 1-day trips have more flying time per day, on average, and thus the pilot living near the domicile will work fewer days to accumulate the 80 flight hours for the month; the pilot with a long distance commute will have more work days and fewer days off to accumulate the same number of flight hours.

Airline Route Networks and Crew Basing

The point from which a pilot begins duty (at his or her base of operations, or domicile) is influenced by airline management practices that vary within the industry. For example, many scheduled airlines—those that operate on specific routes at scheduled times—operate a hub-and-spoke route network in which many flights converge on one airport (the hub) at about the same time so that passengers will have an opportunity to connect conveniently to a flight that is going to their ultimate destination (a spoke). Either a hub or a spoke city could be a pilot's domicile.

Basing pilots at a hub can be attractive for airlines from the point of view of scheduling flexibility and for exchanging crews during connecting operations in the midst of an operating day. Even in a hub-and-spoke system, though, many airplanes are positioned at the spoke airport locations overnight, and basing pilots at a spoke airport can reduce the expenses of providing overnight accommodations ("overnighting") for the pilots who work the originating and terminating flights of the day. In any case, the scheduling and routing of crews does not have to match that of the aircraft. For airlines using domicile basing, whether located at a hub, spoke, or elsewhere, the airlines typically leave the pilots responsible for performing the commute—by whatever modes and means necessary—so as to be at the domicile reliably on time and ready for duty.

In contrast to the practices of most major scheduled airlines, some airlines, often those offering mostly nonscheduled service,[4] have crew-basing practices that reflect shared commuting responsibility among the company and pilots. These practices include home basing, in which the airline arranges a reserved seat on a flight from the pilot's home to the city from which the pilot's flight departs. Another practice is gateway basing, in which the airline arranges a flight (when necessary) from a specified gateway city to the departure city of the pilot's first flight, and the pilot is responsible for the commute from home to the gateway city. For both home basing and gateway basing, a hotel may be provided to ensure a sleep opportunity prior to the first flight.

Competitive and External Factors

The evolving structure of the airline industry also affects the environment in which pilots make commuting decisions. Some airlines' responses to a changing competitive environment have involved establishing new hubs and downsizing or closing existing hubs, as well as starting

[4]Nonscheduled airlines operate on customer demand without a regular schedule.

service to cities they previously did not serve or ending service to some cities. Airline mergers and acquisitions have also led to downsizing or elimination of hubs believed to be redundant in the post-merger route structure.

These sorts of changes may lead to domicile expansions, contractions, closings, or openings, with changes to where a pilot is domiciled. Changes to domiciles are handled, typically, through seniority-based bidding: pilots with relatively less seniority may sometimes be involuntarily moved to new domiciles in other parts of the United States (or even other parts of the world), or, in the extreme, furloughed from the company. Subsequently, recalls from furloughs in response to increases in travel demand may result in pilots being recalled to a domicile that is different from the domicile from which they were released. Seasonal scheduling causes other complications: it may result in changes in pilot domiciles in order to accommodate increased or decreased passenger demand for particular routes. In the absence of pilot commuting, these changes in the airline operating environment could lead to large-scale, sometimes short-term, relocations of pilots and families or inflexibility in the airlines' ability to adjust to changes in staffing needs. The practice of pilot commuting thus enables airlines to adjust to these changes more readily, typically without incurring pilot relocation costs to a new domicile.

Pilot Preferences

The practice of pilot commuting holds benefits not only for the airlines, but also for pilots. Pilots commute to some extent because they can and to some extent because they want to—how and how far a pilot chooses to commute depends on a host of personal and professional decisions involving family, economics, and logistics. Various combinations of work schedules, travel time, and ability to commute are feasible and, in the eyes of the person undertaking the commute, preferable to not commuting.

Pilots who have provided input for this study to date have told the committee that they commute because of both economic and life-style considerations. A pilot may choose a community of residence because of cost of living or tax advantage rather than living in his or her assigned domicile. A community may be selected based on such quality-of-life factors as a desired geographic region, proximity to a school system, or the existence of a support infrastructure for family while the pilot is on extended flight duty. Commuting also enables a pilot to maintain a stable residence if he or she is reassigned to another domicile.

Although any of the cities in a company's hub-and-spoke network could be a given pilot's domicile, there is no guarantee that the company will maintain a given hub-and-spoke pattern long enough for pilots to be assured that their choice for a permanent domicile will remain part of their company's hub-and-spoke network. A point frequently made to the committee was that commuting can provide pilots and their families an aspect of certainty and control when facing the likelihood of mergers, domicile changes, furloughs, and the like, even when considering these as potential disruptions in the future.

Airline Policies and Practices Related to Commuting

Various airline policies and practices may facilitate or hinder commuting and affect pilots' decisions regarding commuting. These include policies related to the consequences of failing to report to the domicile on time because of commuting, as well as those related to sick leave and fatigue; the practices include the ease of commuting (e.g., being able to reserve a passenger seat or jump seat for the commute) and the opportunities for rest (e.g., in-base rest facilities). For the most part, policies directly related to commuting are unregulated and subject to collective bargaining agreements. The committee has requested information from airlines, airline associations, and pilot associations about these policies and practices.

6
SLEEP, CIRCADIAN RHYTHMS, AND FATIGUE

Fatigue

Over the past several decades, the scientific knowledge base about the causes of fatigue and its effects on performance has grown significantly. The FAA-supported *Aviation Fatigue Management Symposium: Partnerships for Solutions* included several presentations summarizing the state of the science relevant to fatigue in aviation (and other transportation modes) (Federal Aviation Administration, 2008). Additional work was presented in the 2009 *International Conference on Fatigue Management in Transportation Operations: a Framework for Progress* (United States Department of Transportation, 2009).

It is clear that fatigue has multiple interactive sources. The primary ones that may be relevant to pilots' commutes include duration of time awake prior to work, duration of time slept prior to work, restfulness of sleep (i.e., sleep continuity) prior to work, and the biological time (i.e., circadian phase) at which commuting occurs relative to the start of work. The duration of time at work (i.e., time on task) is a regulated factor for fatigue mitigation.

In the aviation industry, commutes that involve travel across multiple time zones have the potential to exacerbate the fatigue associated with commuting, as can chronic restriction of sleep for multiple days prior to commuting. It is important to recognize that these fatigue effects can be mitigated to some extent by following good sleep hygiene practices[5] in the period between the end of the commute and the time of reporting for duty. It is unclear at this point, however, the extent to which such practices are followed by commuting pilots.

Extensive scientific evidence documents the multiple negative effects of fatigue on performance for tasks that are similar to those required to operate a commercial aircraft. These include adverse effects from fatigue on alertness and vigilant attention, on the speed and accuracy of performing tasks, on working memory, and on higher cognitive functions such as decision-making. The Institute of Medicine (IOM) defines fatigue as "an unsafe condition that can occur relative to the timing and duration of work and sleep opportunities" (Institute of Medicine, 2009, p. 218). It further states:

> In healthy individuals, fatigue is a general term used to describe feelings of tiredness, reduced energy, and the increased effort needed to perform tasks effectively and avoid errors. It occurs as performance demands increase because of work intensity and work duration, but it is also a product of the quantity and quality of sleep and the time of day work occurs. (Institute of Medicine, 2009, p 218, drawing on Dinges [2001]).

---

[5]Good sleep hygiene practices generally refer to those behaviors that effectively control all behavioral and environmental factors that precede sleep and may interfere with sleep, to ensure the sleep is as restful as possible, in order to promote daytime alertness, or help treat or avoid certain sleep disorders (Thorpy, 2011).

The extent to which commuting may contribute to pilot fatigue at work—by reducing sleep time, extending wake time prior to duty, or interrupting a habitual nocturnal sleep period—is not known. Moreover, pilot commuting practices and individual day-to-day experiences are characterized by tremendous variability.

Sleep and Circadian Rhythms

Further complicating an understanding of the relationship between commuting and pilot fatigue are inadequate data on the timing, duration, and quality of sleep before and during commutes. "Quality of sleep" encompasses factors that can affect the recuperative value of sleep, immediately prior to and during a commute period, such as noise, light, body posture, and ambient temperature.

Scientific understanding of sleep physiology is fundamental to the science of fatigue. Humans spend approximately one-third of their lives asleep. Circadian rhythms are daily rhythms in physiology and behavior that control the timing of the sleep/wake cycle and influence physical and cognitive performance, activity, food consumption, body temperature, heart rate, muscle tone, and some aspects of hormone secretion. When an individual remains awake into his or her habitual nocturnal sleep period, acute sleep loss (time awake extending beyond 16-18 hours) develops: it is characterized by a natural, physiological pressure to sleep (Institute of Medicine, 2009; Van Dongen and Dinges, 2005). This elevated homeostatic sleep drive in the human brain due to being awake too long creates a high pressure for sleep even during daytime work, increasing subjective fatigue and sleepiness while decreasing simple and complex attention and working memory, as well as other cognitive performance functions. These changes can result in adverse effects on performance that can be especially problematic when time awake while working is beyond 16 hours, when sleep prior to work is below 6 hours, and when work is being undertaken at a time when the body is biologically programmed to be asleep (i.e., an individual's habitual nocturnal sleep period), which is most often between 10:00 p.m. and 7:00 a.m. (Basner and Dinges, 2009; Institute of Medicine, 2009; Van Dongen and Dinges, 2005).

Fatigue-related performance deficits from inadequate sleep can vary markedly across a day and while being awake at night (without sleep) because these two factors are not additive. Rather, sleep and circadian drives in the brain interact nonlinearly in the control of performance and alertness (Dijk et al., 1992; Goel et al., 2011). For example, during a 48-hour period of continuous wakefulness, there is a peak in poor performance after 24-28 hours of being awake (e.g., between 6:00 and 10:00 a.m. the following day). Subsequently, performance impairments from a night without sleep are actually somewhat less by 6:00-10:00 p.m. the following day (i.e., at 36-40 hours of being awake) relative to the peak for poor performance occurring earlier that morning (Goel et al., 2011). The detrimental effects of fatigue on performance may be exacerbated by a tendency for individuals to have reduced awareness of the cognitive performance deficits that result, even as these deficits increase in frequency with consecutive days of inadequate sleep (Van Dongen et al., 2003a).

Although the effects of acute sleep deprivation on performance may be influenced by many factors—often referred to as masking factors (Goel et al., 2011)—a recent meta-analysis of 70 articles that covered 147 cognitive tests of several moderators identified time awake as the most significant predictor of behavior during a period of acute sleep deprivation (Lim and Dinges, 2010). This finding could be especially relevant for pilots who get little to no sleep the day before a flight and then undertake a lengthy duty day.

Much is known about the cognitive and functional deficits that result when healthy adult volunteers remain awake for 24-40 hours (Goel et al., 2009b; Harrison and Horne, 2000; Institute of Medicine, 2006, 2009; Philibert, 2005). That scientific understanding of the effects of sleep deprivation on cognitive functions has accumulated for more than a century (for reviews of this extensive literature see Dinges and Kribbs, 1991; Durmer and Dinges, 2005; Harrison and Horne, 2000; Institute of Medicine, 2009; Kleitman, 1963; Patrick and Gilbert, 1896). Additionally, recent advances in neuroimaging technologies have provided further insights into physiological changes in the brain and underlying performance functions that manifest themselves when fatigue results from reduced sleep (Bell-McGinty et al., 2004; Chee and Choo, 2004; Chee et al., 2006, 2008; Chuah et al., 2006; Drummond et al., 1999, 2000, 2005; Habeck et al., 2004; Institute of Medicine, 2009; Lim et al., 2007; Lim et al., 2010; Portas et al., 1998; Thomas et al., 2000; Wu et al., 2006).

It is now recognized that although most adults exposed to a night without sleep experience fatigue related declines in performance, the timing and severity of the declines vary across individuals, including pilots (Doran et al., 2001; Leproult et al., 2003; Van Dongen et al., 2004, Institute of Medicine, 2009). These differences in individual cognitive vulnerability to sleep loss may have a basis in genes regulating sleep and circadian rhythms (Institute of Medicien, 2009; Goel et al., 2009a, 2010). Persons with untreated sleep disorders are also subject to individual vulnerability and may experience negative effects on their performance and safety beyond those experienced by healthy individuals.

In addition to acute sleep deprivation, fatigue can be exacerbated by chronic partial sleep loss, also known as cumulative sleep debt, which occurs when the sleep obtained over multiple days is too short in duration to maintain behavioral alertness during the daytime (Van Dongen et al., 2003b). There is extensive scientific evidence that chronic undersleeping results in cumulative performance deficits across days and that the rate of the performance decline is inversely proportional to the sleep obtained (Belenky et al., 2003; Dinges et al., 1997; Van Dongen et al., 2003a). The performance deficits from chronic sleep restriction can also accumulate across days to levels equivalent to those found after one and even two nights without any sleep (Van Dongen et al., 2003a). Chronic sleep restriction that is followed by a night of little to no sleep results in severe deficits in cognitive performance (Banks et al., 2010). The threshold at which chronic sleep restriction appears to adversely affect performance in a majority of healthy adults is when time in bed for sleep is at 6 hours or less per 24 hours on a consistent basis. It is important to note that this threshold is the case for the vast majority of people who habitually need 7-8 hours sleep a night, but that the studies were not done on the minority of adults who are naturally short sleepers, requiring 6 or fewer hours per night. It appears that recovery from chronic sleep loss often requires extended periods of sleep (Banks et al., 2010; Belenky et al., 2003).

Since pilot commuting can involve sleep opportunities while seated or semirecumbent (e.g., in a car, bus, or plane), the recovery potential of such sleep becomes an important issue to understand in fatigue mitigation. For instance, both the angle of the back while sleeping and the characteristics of potential napping opportunities influence the overall restfulness of sleep. Sleeping in environments not conducive to sleep can result in reduced recovery potential, even during naps (Dinges et al., 1981). Sleeping in an upright position results in reduced sleep quality (i.e., less sleep time with increased awakenings) in comparison with sleeping in a lying flat or reclined position (Dinges et al., 1981; Nicholson and Stone, 1987).

Although napping has been shown to be an effective technique for restoring alertness and performance during periods of continued wakefulness, it is the timing and length of a nap, along with the timing of the nap in the circadian phase, that moderates the benefits of napping for performance (Bonnet, 1991; Caldwell et al. 2009; Dinges et al., 1987; Matsumoto and Harada, 1994; Rogers et al., 1989; Rosa, 1993; Vgontzas et al., 2007; Webb, 1987).

Recognition of the complex nature of the multiple interacting factors that influence the build up and reduction of fatigue as a state that can affect performance has been at the core of the development and application of various fatigue management strategies. The science of fatigue management has developed rapidly over the past decade in civilian transportation sectors, with much of the applied research sponsored originally by the military, where sustained and continuous operations pose acute fatigue-related challenges. There are now several well-documented candidate systems for measuring fatigue and its negative effects on performance. There are also mathematical models that demonstrate limited ability to predict fatigue using information on duty time and scheduling, sleep quantity and quality, circadian and time-zone information, and other variables.

There have been steady advances in various fatigue management technologies, including devices that monitor an operator's level of alertness or performance, as well as devices that predict fatigue in advance of a work cycle or trip (Balkin et al., 2011). Although some of these technological approaches to fatigue management show considerable promise, there remain important unresolved questions and limitations regarding the validity and reliability of their use and acceptance by operators and industries (Balkin et al., 2011).

Among the most popular technologies for assisting in fatigue management are the mathematical models derived from research on the dynamics of performance relative to the interactions of sleep duration, wake duration, and circadian phase, which claim to predict performance during different work-rest schedules. A workshop sponsored by the U.S. Department of Defense, the U.S. Department of Transportation, and the National Aeronautics and Space Administration provided an opportunity to conduct an initial evaluation and comparison of seven of these mathematical models (Mallis et al., 2004). Although predictions of performance were promising, the evaluation showed that further research was needed to demonstrate the models' validity and reliability using real-world data and that the models could not make reliable predictions of group performance risks from fatigue over multiday schedules (Dinges, 2004; Van Dongen, 2004). There is considerable research now under way to address how to use these measures, models, and other knowledge in the design and implementation of staffing and work-scheduling programs in order to minimize fatigue (see National Research Council, 2007). The potential for practical application of these models in the commercial aviation context—and particularly in relation to pilot commuting—is unclear at this time.

The issue of fatigue in safety-sensitive work operations cuts across many industries and has been addressed broadly in the scientific literature. The combination of work demands, sleep restriction, and circadian factors can negatively affect alertness, performance, speed, accuracy, and central nervous system functioning (Cabon et al., 1993; Goel et al, 2009b): see Box 4.

BOX 4
Factors in the Risk of Fatigue-Related Errors and Accidents

Risks of fatigue-related errors and accidents stem from multiple interrelated and interacting aspects of work, rest, and sleep. These include but are not limited to

(1) duration of work periods within a single day and over time,
(2) time of day at which work occurs,
(3) variation in the timing of work within and between weeks,
(4) duration of sleep obtained on work days and on non-work days,
(5) frequency and duration of days off from work,
(6) different vulnerabilities of workers to fatigue from these factors, and
(7) volume and intensity of work.

SOURCE: Institute of Medicine (2009, pp. 218-219) citing the works of Dinges (1995), Drake et al., (2004), Folkard et al. (2005), Rosa (2001), and Van Dongen (2006).

7
CURRENT REGULATORY PROCESS

Current federal flight duty time regulations (14 CFR 91 and 121) do not address pilot commuting. There is only a general requirement (in Part 91.13) that crew members should not be careless or reckless in the operation of an aircraft.

In response to P.L. 111-216 (the Airline Safety and Federal Aviation Administration Extension Act of 2010), the FAA's current proposed regulation related to flight and duty time attempts to take advantage of the available research on fatigue, sleep, and circadian rhythms and, among other things, to consider the effects of commuting, means of commuting, and the length of the commute on fitness for duty. In these proposed regulations, time spent commuting is not considered duty time.

As noted above, as part of its effort to update these regulations the FAA issued an NPRM in September 2010 describing regulatory revisions in which commuting is included. The proposed regulations present commuting as fundamentally an issue of fitness for duty, defining a responsible commuter as a pilot who (U.S. Department of Transportation, 2010, NPRM, p. 55,874):

plans his or her commute to minimize its impact on his or her ability to get meaningful rest shortly before flying, thus fulfilling the proposed requirement that he or she reports for an FDP [flight duty period] rested and prepared to perform his or her assigned duty.

In the NPRM (U.S. Department of Transportation, 2010 p. 55,875) the FAA states that "It is inappropriate to rely simply on the requirements to report 'fit for duty' in Part 91"[6] and proposes a new Part 117 specific to fitness for duty. As a complement to issuance of the NPRM, the FAA issued a draft advisory circular (AC 120-FIT) on fitness for duty in which fitness for duty is considered as a joint responsibility of the air carrier and the crew member and outlines specific aspects of their responsibilities.

The NPRM has raised concerns about regulating commuting related both to infringement of personal choice and the possible inadvertent effect that would prompt "irresponsible commuting." Hence, the proposed regulation points out that commute time should not be considered rest and that carriers have an obligation to "consider the commuting times required by individual flight crew members to ensure they can reach their home base while still receiving the required opportunity for rest." It also conveys the FAA's view that "irresponsible commuting" results primarily from a lack of pilot education regarding what activities are fatiguing and how to mitigate becoming fatigued. Pilot education is one of the specified objectives of the draft advisory circular on fitness-for-duty mentioned above. The effect of commuting on fatigue is also one element of a recommended training curriculum specified in the NPRM.

This committee has approached many international regulatory and safety oversight organizations, operators, and pilot associations—including the International Civil Aviation Organization (ICAO), International Air Transport Association, Flight Safety Foundation, and International Federation of Airline Pilots' Association to obtain information regarding existing regulations, policies, and best practices regarding commuting outside the United States. In addition to regulatory approaches to fatigue, new developments both in the science of fatigue and performance and in management and regulatory philosophies have led to another approach in the transportation domain, usually termed fatigue risk management systems. These systems are focused on integrating scientific knowledge about fatigue and its management with the realities of airline operations. In essence, these systems recognize that responsibility for managing fatigue-related safety risks is a shared responsibility of regulatory authorities, operators, and individual pilots.

Fatigue risk management systems are currently under development by several airlines, and industry and professional groups, as well as national and international regulatory agencies, are involved in research and development efforts. The ICAO established a fatigue risk management systems task force to review scientific and operational knowledge and to develop detailed regulatory standards and guidance for member countries on implementation of such systems (see International Civil Aviation Organization, 2009). The proposed Standard and Recommended Practice was approved by the ICAO's Air Navigation Committee on December 14, 2010; if it is approved by the ICAO's council at its annual meeting in March 2011, it will be effective in July 2011.

P.L. 111-216 required U.S. airlines to submit to the FAA drafts of their Fatigue Risk Management Plans. Although these FRM Plans do not yet correspond to ICAO's Standard and Recommended Practice, as part of its data gathering, the committee has requested information

---

[6]Part 91 does not specifically refer to "fitness for duty;" rather, as noted above it states that no person can operate the aircraft in a "careless and reckless manner."

from U.S. airlines on whether their fatigue risk management plans take commuting into account and the committee will discuss the results in its final report.

The FAA approach is compatible with ICAO's fatigue risk management systems initiative and the trend over the past two decades of many U.S. federal regulatory agencies to shift more responsibility to the organizations they regulate and to encourage cooperative rather than adversarial relationships. Generally, these initiatives rely on management systems using continuous monitoring to identify and mitigate potential risks before they have safety consequences Such voluntary FAA programs include the Aviation Safety Action Program and the Flight Operational Quality Assurance Program.

The Occupational Safety and Health Administration (OSHA) also has several voluntary compliance strategies that take a similar approach.[7] (The Food and Drug Administration's Hazard Analysis and Critical Control Points Program for food safety is another management systems approach. The success of such programs is a matter of some disagreement and, as noted by the U.S. Government Accountability Office (2004) in the case of OSHA, rigorous evaluation is needed to examine their effectiveness.

8
NEXT STEPS

There are many issues that complicate consideration of whether and how commuting affects pilot fatigue in a manner detrimental to flight safety, not the least of which is the lack of comprehensive, industrywide data on the prevalence and characteristics of commuting. On the basis of the comments and documents the committee has reviewed to date, many airlines and pilots believe that pilot fatigue is a safety concern. However, the extent and circumstances under which commuting contributes to fatigue remain unclear. Airline policies and practices, characteristics of the aviation system, and individual pilot behavior all play a role in pilot fatigue. It seems to the committee that it is important to note that safety in scheduled air transportation has continued to improve over time, to the point where catastrophic, fatal accidents in such operations are statistically rare events. Although much remains to be done in the way of data collection and analysis, pilot commuting appears to be a fairly widespread aspect of these operations.

Over the next several months, the committee will follow up on its requests for information, continue to review relevant literature and information received, and attempt to analyze the role of the many factors involved in the issue of pilot commuting and fatigue. The committee's final report, in keeping with the charge, will outline its thoughts on potential next steps, possibly including promising practices, recommended changes to FAA regulations, administrative actions, and research priorities.

[7]For details of OSHA's cooperative programs, see http://www.osha.gov/dcsp/compliance_assistance/index_programs.html [January 2011].

REFERENCES

Balkin, T.J., W.J. Horrey, R.C. Graeber, C.A. Czeisler, and D.F. Dinges. (2011). The challenges and opportunities of technological approaches to fatigue management. *Accident Analysis and Prevention* 43(2):565-572.

Banks, S., H.P.A. Van Dongen, G. Maislin, and D.F. Dinges. (2010). Neurobehavioral dynamics following chronic sleep restriction: Dose-response effects of one night for recovery. *Sleep* 33(8):1,013-1,026.

Basner, M., and D.F. Dinges. (2009). Dubious bargain: Trading sleep for Leno and Letterman. *Sleep* 32(6):747-752.

Belenky, G., N.J. Wesensten, D.R. Thorne, M.L. Thomas, H.C. Sing, D.P. Redmond, M.B. Russo, and T.J. Balkin. (2003). Patterns of performance degradation and restoration during sleep restriction and subsequent recovery: A sleep dose-response study. *Journal of Sleep Research* 12:1-12.

Bell-McGinty, S., C. Habeck, H.J. Hilton, B. Rakitin, N. Scarmeas, E. Zarahn, J. Flynn, R. DeLaPaz, R. Basner, and Y. Stern. (2004). Identification and differential vulnerability of a neural network in sleep deprivation. *Cerebral Cortex* 14(5):496-502.

Bonnet, M.H. (1991). The effect of varying prophylactic naps on performance, alertness and mood throughout a 52-hour continuous operation. *Sleep* 14(4):307-315.

Cabon P.H., A. Coblentz, R. Mollard, and J. P. Fouillot. (1993). Human vigilance in railway and long-haul flight operation. *Ergonomics* 36(9):1,019-1,033

Caldwell, J.A., M.M. Allis, J.L. Caldwell, M.A. Paul, J.C. Miller, and D.F. Neri. (2009). Aerospace Medical Association aerospace fatigue countermeasures. Subcommittee of the Human Factors Committee, Fatigue Countermeasures in Aviation. *Aviation, Space, and Environmental Medicine* 80:29-59.

Chee, M.W.L., and W.C. Choo. (2004). Functional imaging of working memory after 24 hr of total sleep deprivation. *Journal of Neuroscience* 24(19):4,560-4,567.

Chee, M.W.L., L.Y.M. Chuah, V. Venkatraman, W.Y. Chan, P. Philip, and D.F. Dinges. (2006). Functional imaging of working memory following normal sleep and after 24 and 35 h of sleep deprivation: Correlations of fronto-parietal activation with performance. *Neuroimage* 31(1):419-428.

Chuah, L.Y.M., V. Venkatraman, D.F. Dinges, and M.W.L. Chee. (2006). The neural basis of inter-individual variability in inhibitory efficiency following sleep deprivation. *Journal of Neuroscience* 26(27):7,156-7,162.

Dijk, D.J., J.F. Duffy, and C.A. Czeisler. (1992). Circadian and sleep/wake dependent aspects of subjective alertness and cognitive performance. *Journal of Sleep Research* 1(2):112-117.

Dinges, D.F. (1995). An overview of sleepiness and accidents. *Journal of Sleep Research,* Supplement 4(2):4-14.

Dinges, D.F. (2001). Stress, fatigue, and behavioral energy. *Nutrition Reviews* 59(1 Pt II):S30-S32.

Dinges, D.F. (2004). Critical research issues in development of biomathematical models of fatigue and performance. *Aviation Space and Environmental Medicine* 75(3):A181.

Dinges, D.F., and N.B. Kribbs. (1991). Performing while sleepy: Effects of experimentally induced sleepiness. In T. Monk (Ed.), *Sleep, Sleepiness and Performance*, pp. 97-128. Chichester, England: John Wiley & Sons.

Dinges, D.F., E.C. Orne, F.J. Evans, and M.T. Orne. (1981). Performance afternaps in sleep conducive and alerting environments. In L.C. Johnson, D.I. Tepas, W.P. Colquhoun, and M.J. Colligan (Eds.), *Biological Rhythms, Sleep and Shift Work. Advances in Sleep Research* (Vol. 7), pp. 539-552. New York: Spectrum Publications.

Dinges, D.F., M.T. Orne, W.G. Whitehouse, and E.C. Orne. (1987). Temporal placement of a nap for alertness: Contributions of circadian phase and prior wakefulness. *Sleep* 10(4):313-329.

Dinges, D.F., F. Pack, K. Williams, K.A. Gillen, J.W. Powell, G.E. Ott, C. Aptowicz, and A.I. Pack. (1997). Cumulative sleepiness, mood disturbance, and psychomotor vigilance performance decrements during a week of sleep restricted to 4-5 hours per night. *Sleep* 20(4):267-277.

Doran, S.M., H.P.A. Van Dongen, and D.F. Dinges. (2001). Sustained attention performance during sleep deprivation: Evidence of state instability. *Archives Italiennes de Biologie* 139(3):253-267

Drake, C.L., T. Roehrs, G. Richardson, J.K. Walsh, and T. Roth. (2004). Shift work sleep disorder: Prevalence and consequences beyond that of symptomatic day workers. *Sleep* 27(8):1,453-1,462.

Drummond, S.P.A., G.G. Brown, J.L. Stricker, R.B. Buxton, E.C. Wong, and J.C. Gillin. (1999). Sleep deprivation-induced reduction in cortical functional response to serial subtraction. *NeuroReport* 10:3,745-3,748.

Drummond, S.P., G.G. Brown, J.C. Gillin, J.L. Stricker, E.C. Wong, and R.B. Buxton. (2000). Altered brain response to verbal learning following sleep deprivation. *Nature* 403(6770):655-657.

Drummond, S.P.A., A. Bischoff-Grethe, D.F. Dinges, L. Ayalon, S.C. Mednick, and M.J. Meloy. (2005). The neural basis of the psychomotor vigilance task. *Sleep* 28(9):1,059-1,068.

Durmer, J.S., and D.F. Dinges. 2005. Neurocognitive consequences of sleep deprivation. *Seminars in Neurology* 25(1):117-129.

Federal Aviation Administration. (2008). *Proceedings: Aviation Fatigue Management Symposium: Partnerships for Solutions*. Washington, DC: Federal Aviation Administration.

Folkard, S., D.A. Lombardi, and P.T. Tucker. (2005). Shiftwork: Safety, sleepiness and sleep. *Industrial Health* 43(1):201-223.

Goel, N., S. Banks, E. Mignot, and D.F. Dinges (2009a). PER3 polymorphism predicts cumulative sleep homeostatic but not neurobehavioral changes to chronic partial sleep deprivation. *PLoS ONE* 4(6):e5874

Goel, N., H. Rao, J.S. Durmer, and D.F.Dinges. (2009b). Neurocognitive consequences of sleep deprivation. *Seminars in Neurology* 29(4):320-339.

Goel, N., S. Banks, E. Mignot, D.F. Dinges (2010). DQB1 *0602 predicts interindividual differences in physiological sleep, sleepiness and fatigue. *Neurology* 75:1,509-1,519.

Goel, N., H.P.A. Van Dongen, D.F. Dinges. (2011). Circadian rhythm in sleepiness, alertness and performance. In M.H. Kryger, T. Roth, W.C. Dement (Eds.). *Principles and Practice of Sleep Medicine* (5th edition), pp. 4,445-4,455. Philadelphia, PA: Elsevier.

Habeck, C., B.C. Rakitin, J. Moeller, N. Scarmeas, E. Zarahn, T. Brown, and Y. Stern. (2004). An event-related fMRI study of the neurobehavioral impact of sleep deprivation on performance of a delayed-match-to-sample task. *Cognitive Brain Research* 18(3):306-321.

Harrison, Y., and J.A. Horne (2000). The impact of sleep deprivation on decision making: A review. *Journal of Experimental Psychology: Applied* 6(3):236-249.
Institute of Medicine. (2006). *Sleep Disorders and Sleep Deprivation: An Unmet Public Health Problem.* Committee on Sleep Medicine and Research, H.R. Colten and B.M. Altevogt, Eds. Board on Health Sciences Policy. Washington, DC: The National Academies Press.
Institute of Medicine. (2009). *Resident Duty Hours: Enhancing Sleep, Supervision, and Safety.* Committee on Optimizing Graduate Medical Trainee (Resident) Hours and Work Schedule to Improve Patient Safety, C. Ulmer, D.M. Wolman, M.M.E. Johns, Eds. Washington, DC: The National Academies Press.
International Civil Aviation Organization. (2009). *The ICAO Fatigue Risk Management Systems Task Force*. Presentation by Mitchell A. Fox, Chief Flight Safety Section, ICAO. Available: http://www.icao.int/nacc/meetings/2009/RASGPA02/Pres/Day3/3-3/rasg-pa_2%20fatigue.pdf [August 2010].
Kleitman, N. (1963). *Sleep and Wakefulness*. Chicago, IL: University of Chicago Press
Krueger, G. (2004). *Technologies and Methods for Monitoring Driver Alertness and Detecting Driver Fatigue: A Review Applicable To Long-Haul Truck Driving*. Technical Report for American Transportation Research Institute and Federal Motor Carrier Safety Administration.
Leproult, R., E.F. Colecchia, A.M. Berardi, R. Stickgold, S.M. Kosslyn, and E. Van Cauter. (2003). Individual differences in subjective and objective alertness during sleep deprivation are stable and unrelated. *American Journal of Physiology-Regulatory Integrative and Comparative Physiology* 284(2):R280-R290.
Lim, J., and D.F. Dinges. (2010) A meta-analysis of the impact of short-term sleep deprivation on cognitive variables. *Psychological Bulletin* 136(3):375-389.
Lim, J., W.C. Choo, and M.W.L. Chee. (2007). Reproducibility of changes in behaviour and fMRI activation associated with sleep deprivation in a working memory task. *Sleep* 30(1):61-70
Lim, J., J.C. Tan, S. Parimal, D.F. Dinges, and M.W.L. Chee (2010). Sleep deprivation impairs object-selective attention: A view from the ventral visual cortex. *PLoS ONE* 5(2):e9087
Mallis, M.M., S. Mejdal, T.T. Nguyen, and D.F. Dinges (2004). Summary of features of seven biomathematical models of human fatigue and performance. *Aviation Space and Environmental Medicine* 75(3):A4.
Matsumoto, K., and M. Harada. (1994). The effect of night-time naps on recovery from fatigue following night work. *Ergonomics* 37(5):899-907.
National Transportation Safety Board. (1994). *A Review of Flightcrew-Involved, Major Accidents of U.S. Air Carriers, 1978 Through 1990.* Safety Study NTSB/SS-94/01. Available: http://libraryonline.erau.edu/online-full-text/nstb/safety-studies/5594-01.pdf [September 2010].
National Transportation Safety Board. (2010). *Loss of Control on Approach, Colgan Air, Inc. Operating as Continental Connection Flight 3407, Bombardier DHC-8-400, N200WQ, Clarence Center, New York, February 12, 2009.* Accident Report NTSB/AAR-10/01, PB2010-910401. Washington, DC: National Transportation Safety Board.
Nicholson, A.N., and B.M. Stone. (1987). Influence of back angle on the quality of sleep in seats. *Ergonomics* 30(7):1,033-1,041.
Patrick, G.T.W., and J.A. Gilbert. (1896). On the effects of loss of sleep. *Psychological Review* 3(5):469-483.

Philibert, I. (2005). Sleep loss and performance in residents and nonphysicians: A meta-analytic examination. *Sleep* 28(11):1,392-1,402.

Portas, C.M., G. Rees, A.M. Howseman, O. Josephs, R. Turner, and C.D. Frith. (1998). A specific role for the thalamus in mediating the interaction of attention and arousal in humans. *Journal of Neuroscience* 18(21): 8,979-8,989.

Portas, C.M., B. Bjorvatn, S. Fagerland, J. GrØnli, V. Mundal, E. SØrensen, et al. (1998). On-line detection of extracellular levels of serotonin in dorsal raphe nucleus and frontal cortex over the sleep/wake cycle in the freely moving rat. *Neuroscience* 83(3):807-814.

Public Law 111-216. Enacted August 1, 2010. Airline Safety and Federal Aviation Administration Extension Act of 2010. Available: http://frwebgate.access.gpo.gov/cgibin/getdoc.cgi?dbname=111_cong_bills&docid=f:h5900rds.txt.pdf [January 2011].

Rogers, A.S., M.B. Spencer, B.M. Stone, and A.N. Nicholson. (1989). The influence of a 1 h nap on performance overnight. *Ergonom*ics 32 (10):1,193-1,205.

Rosa, R.R. (1993). Napping at home and alertness on the job in rotating shift workers. *Sleep* 16(8):727-735.

Rosa, R.R. (2001). Examining work schedules for fatigue: It's not just hours of work. In P.A. Hancock and P.A. Desmond (Eds.), *Stress, Workload, and Fatigue*, pp. 513-528. Mahwah, NJ: Lawrence Erlbaum Associates, Inc.

Thomas, M., H. Sing, G. Belenky, H. Holcomb, H. Mayberg, R. Dannals, H. Wagner, D. Thorne, K. Popp, L. Rowland, A. Welsh, S. Balwinski, and D. Redmond. (2000). Neural basis of alertness and cognitive performance impairments during sleepiness. I. Effects of 24 h of sleep deprivation on waking human regional brain activity. *Journal of Sleep Research* 9(4):335-352.

Thorpy, M.J. (2011) Circadian rhythm in sleepiness, alertness and performance. In M.H. Kryger, T. Roth, and W.C. Dement (Eds.), *Principles and Practice of Sleep* Medicine (5th edition), pp. 680-693. Philadelphia, PA: Elsevier.

United States Department of Transportation. (2009). 2009 International Conference on Fatigue Management in Transportation Operations: a Framework for Progress: Boston, Massachusetts, March 24-26, 2009. United States. Dept. of Transportation.; DOT Human Factors Coordinating Committee (U.S.) 2009 [Washington, DC] Available: http://depts.washington.edu/uwconf/fmto/FatigueManagementAbstracts.pdf [January 2011]

United States Department of Transportation. (2010). Federal Aviation Administration 14 CFR Parts 117 and 121 Flightcrew Member Duty and Rest Requirements: Proposed Rule Federal Register/ Vol.75. No. 177/ Tuesday, September 14, 2010/Proposed Rules. (NPRM).

United States Department of Transportation. Federal Aviation Administration Advisor Circular, Subject: Fitness for Duty AC No: AC 120-FIT.

Van Dongen, H.P.A., G. Maislin, J.M. Mullington, and D.F. Dinges (2003a). The cumulative cost of additional wakefulness: Dose-response effects on neurobehavioral functions and sleep physiology from chronic sleep restriction and total sleep deprivation. *Sleep* 26(2): 117-126.

Van Dongen, H.P.A, N.L. Rogers, and D.F. Dinges. (2003b). Sleep debt: Theoretical and empirical issues. *Sleep & Biological Rhythms* 1(1):5-13.

Van Dongen H.P.A, and D.F..Dinges. (2005) Circadian rhythms in sleepiness, alertness, and performance. In M.H. Kryger, R.Roth, and W.C. Dement (Eds.), *Principles and Practice of Sleep Medicine*, pp. 435-443. Philadelphia, PA: Elsevier.

Van Dongen, H.P.A. (2006). Shift work and inter-individual differences in sleep and sleepiness. *Chronobiology International* 23(6):1,139-1,147.

Van Dongen, H.P., M.D. Baynard, G. Maislin, and D.F. Dinges. (2004). Systematic interindividual differences in neurobehavioral impairment from sleep loss: Evidence of trait-like differential vulnerability. *Sleep* 27(3):423-433.

Vgontzas A.N., S. Pejovic, E. Zoumakis, H.M. Lin, E.O. Bixler, M. Basta, J. Fang, A. Sarrigiannidis, G.P. Chrousos, et al. (2007). Daytime napping after a night of sleep loss decreases sleepiness, improves performance, and causes beneficial changes in cortisol and interleukin-6 secretion. *American Journal of Physiology-Endocrinology and Metabolism* 292(1):E253-E261.

Webb, W.B. (1987). The proximal effects of two and four hour naps within extended performance without sleep. *Psychophysiology* 24(4):426-429.

Wu, J.C., J.C. Gillin, M.S. Buchsbaum, P. Chen, D.B. Keator, N.K. Wu, L.A. Darnall, J.H. Fallon, and W.E. Bunney. (2006). Frontal lobe metabolic decreases with sleep deprivation not totally reversed by recovery sleep. *Neuropsychopharmacology* 31(12):2,783-2,792.

Appendix A
Biographical Sketches of Committee Members and Staff

Clinton V. Oster, Jr. (*Chair*) is a professor at the School of Public and Environmental Affairs at Indiana University. Previously, he served as director of the Transportation Research Center and as associate dean at the School of Public and Environmental Affairs at Indiana University. His research focuses on air traffic management and aviation infrastructure, with an emphasis on aviation safety. His research also includes airline economics, airline competition policy, and energy policy. He has been a consultant to the U.S. Department of Transportation, the Federal Aviation Administration, the National Aeronautics and Space Administration (NASA), the European Bank for Reconstruction and Development, state and local governments, and private-sector companies in the United States, Canada, the United Kingdom, Russia, and Australia. He is a member of the National Aviation Advisory Group of the U.S. Government Accountability Office, and he has been an expert witness for the Environment and Natural Resource Division and the Antitrust Division of the U.S. Department of Justice. He received a B.S.E. in chemical engineering from Princeton University, an M.S. in urban and public affairs from Carnegie Mellon University, and a Ph.D. in economics from Harvard University.

Benjamin A. Berman is a senior research associate in the Human Systems Integration Division at the NASA Ames Research Center (affiliated through San Jose State University) and is a pilot for a major U.S. air carrier with 9,000 hours of flight experience. Before returning to professional flying in 2001, he was on the staff of the National Transportation Safety Board (NTSB), where he served as the chief of the Major Investigations Division and E led the Operational Factors Division (responsible for flight operations, air traffic control, and meteorology investigations), served as the flight operations investigator for major cases including the USAir B-737 accident in Pittsburgh and the ValuJet DC-9 accident in the Everglades, and managed flight crew human factors research projects. He holds an Airline Transport Pilot Certificate with type ratings for the Boeing 777, Boeing 737, Embraer 120, and Dornier 228. He received an A.B. summa cum laude in economics from Harvard College.

J. Lynn Caldwell is a senior research psychologist for the U.S. Air Force Research Laboratory, currently stationed at Wright-Patterson Air Force Base in Ohio. Previously, she was with the U.S. Army's Aeromedical Research Laboratory, where she conducted numerous simulator and in-flight investigations on fatigue countermeasures and circadian rhythms in rated military pilots. She has also been a member of the Warfighter Fatigue Countermeasures Program and a distinguished visiting scholar at the U.S. Air Force Academy. She has served as a fatigue consultant for various U.S. Air Force commands and other military and civilian groups. She frequently provides fatigue management workshops, safety briefings, and training courses to aviation personnel, flight surgeons, commanders, and safety officers. She is certified as a sleep specialist by the American Board of Sleep Medicine. She received a Ph.D. in experimental psychology from the University of Southern Mississippi.

David F. Dinges is a professor and chief of the Division of Sleep and Chronobiology and director of the Unit for Experimental Psychiatry in the Department of Psychiatry and associate director of the Center for Sleep and Respiratory Neurobiology at the University of Pennsylvania School of Medicine. His research focuses on physiological, neurobehavioral, and cognitive effects of sleep loss, disturbances of circadian biology, and stress, and the implications of these unmitigated effects on health and safety. He currently leads the Neurobehavioral and Psychosocial Factors Team for the National Space Biomedical Research Institute. He has been president of the U.S. Sleep Research Society and of the World Federation of Sleep Research and Sleep Medicine Societies, and he has served on the board of directors of the American Academy of Sleep Medicine and the National Sleep Foundation. He is currently editor-in-chief of *Sleep*. His awards include the 2004 Decade of Behavior Research Award from the American Psychological Association and the 2007 NASA Distinguished Public Service Medal. He has an A.B. in psychology from Saint Benedict's College, a M.S. in physiological psychology from Saint Louis University, an honorary M.A. from the University of Pennsylvania, and a Ph.D. in physiological psychology from Saint Louis University.

R. Curtis Graeber is the president of The Graeber Group, Ltd. Previously, he served as the chief engineer for human factors and director of regional safety programs at Boeing Commercial Airplanes and in other several management positions in research, airplane design, and safety. He also led Boeing's efforts to improve regional safety, including industry development and implementation of the global aviation safety roadmap. Before joining Boeing, he led the flight crew fatigue research program at NASA's Ames Research Center and served as chief of flight human factors. He also served as the human factors specialist for the Presidential Commission on the Space Shuttle Challenger Accident. He is a fellow of the Royal Aeronautical Society and the Aerospace Medical Association. He has chaired working groups for the Federal Aviation Administration, the Flight Safety Foundation, and the International Civil Aviation Organization. His safety-related awards include the Guild of Air Pilots and Air Navigators' Cumberbatch Trophy and the Aerospace Medical Association's Boothby-Edwards Award. He serves as chair of Air New Zealand's Independent Alertness Advisory Panel, and he is a member of the Board of Directors of the National Sleep Foundation. He received a Ph.D. in neuropsychology from the University of Virginia.

John K. Lauber is a private consultant. Previously, he served as senior vice president and chief product safety officer for Airbus SAS in Toulouse, France, as vice president of safety and technical affairs for Airbus North America, and as vice president of training and human factors for Airbus Service Company. Prior to joining Airbus he was vice president of corporate safety and compliance at Delta Air Lines. Both Presidents Ronald Reagan and George H.W. Bush appointed him to terms as a member of the National Transportation Safety Board. He has served as chief of the Aeronautical Human Factors Research Office for NASA Ames Research Center, where he was instrumental in the development of advanced flight crew training concepts that are now used by airlines around the world. He is a commercial pilot, with both airplane and helicopter ratings and is type-rated in the B727 and the A320. His numerous awards include NASA's Outstanding Leadership Award and the Boeing/Flight Safety Award for Lifetime Achievement in Aviation Safety He has served as president of the International Federation of Airworthiness and the Association for Aviation Psychology. He holds a Ph.D. degree in neuropsychology from Ohio State University (1969).

David E. Meyer is a faculty member of the Cognition and Cognitive Neuroscience Program in the Department of Psychology at the University of Michigan. Previously, he worked in the Human Information Processing Research Department at the Bell Telephone Laboratories. His teaching and his research have dealt with fundamental aspects of human perception, attention, learning, memory, language, movement production, reaction time, multitasking, executive mental control, human-computer interaction, personality and cognitive style, cognitive aging, cognitive neuroscience, mathematical models, and computational models. He is a fellow in the Society of Experimental Psychologists, the American Psychological Society, the American Psychological Association, and the American Association for the Advancement of Science. The American Psychological Association has honored him with its Distinguished Scientific Contribution Award. He is a member of the National Academy of Sciences. He received a Ph.D. from the University of Michigan.

Mary Ellen O'Connell (*Project Director*) is deputy director for the National Research Council's Board on Human-Systems Integration. At the NRC, she has served as study director for five major consensus studies: on prevention of mental disorders and substance abuse, international education and foreign languages, ethical considerations for research on housing-related health hazards involving children, reducing underage drinking, and assessing and improving children's health. She also organized workshops on welfare reform and children and gun violence. Previously, she held various positions at the U.S. Department of Health and Human Services (HHS), including serving as director of state and local initiatives in the Office of the Assistant Secretary for Planning and Evaluation. Her previous positions also include work on homeless policy and program design at the U.S. Department of Housing and Urban Development and as director of field services for the Commonwealth of Massachusetts. She has a B.A. (with distinction) from Cornell University and a masters in the management of human services from the Heller School for Social Policy and Management at Brandeis University.

Matthew Rizzo is professor of neurology, engineering, and public policy at the University of Iowa. At the university, he is also vice chair for clinical/translational research and director of the division of neuroergonomics, its Visual Function Laboratory, and its instrumented vehicles in the Department of Neurology, as well as director of the University Institute for the Aging Mind. His clinical interests and activities include behavioral neurology, cognitive neuroscience, and memory disorders. His research interests include behavioral disturbances resulting from central nervous system injury, neural substrates of human vision (including attention and visuomotor control), aging and dementia, driving performance, and driving simulation. He has conducted research on fatigue and truckers for the National Institutes of Health and the Iowa Department of Transportation. Dr. Rizzo is a member of the American Academy of Neurology, the American Neurological Association, the Human Factors and Ergonomics Society, the Society for Neuroscience, and the Vision Sciences Society. Dr. Rizzo is a member of the NRC's Board on Human-Systems Integration. He has an M.D. from Johns Hopkins University School of Medicine.

David J. Schroeder is a private consultant. Previously, he was a manager of the Aerospace Human Factors Research Division at the Civil Aero Medical Institute of the Federal Aviation Administration (FAA), where he also served as supervisor of clinical psychology research and was administrator of the FAA's Employee Attitude Survey for 22 years. His research is documented in over 40 Office of Aviation Medicine (OAM) technical reports and in more than 125 presentations in areas of interest such as disorientation, job attitudes, stress, age, shiftwork and fatigue, and color vision. He participated in numerous national human factors working groups and helped develop many international collaborative research projects. He also assisted with the psychological screening of federal air marshals during their post-9/11 hiring increase. He was the Office of Aviation Medicine Manager of the Year in 2005 and led his division to become the OAM Office of the Year in 1999 and 2005. He is past president of the Oklahoma Psychological Association, the APA Division of Applied Experimental and Engineering Psychology, and the Aerospace Medical Association, and was a division representative to the APA Council for three years. He has a Ph.D. in psychology from the University of Oklahoma.

Toby Warden (*Study Director*) is a program officer with the Board on Human-Systems Integration of the National Research Council (NRC). Previously, she worked as a program officer with the NRC's Board on Atmospheric Sciences and Climate of serving as study director for the projects that published *Climate Stabilization Targets: Emissions, Concentrations, and Impacts Over Decades to Millennia* and *When Weather Matters: Science and Service to Meet Critical Societal Needs*. She has nearly a decade's worth of experience as a program manager and community organizer in the fields of public health and youth advocacy in Boston, Massachusetts. Her doctoral research applied quantitative and qualitative methodologies to examine the rise of the U.S. Mayors Climate Protection Agreement. She has a B.A. in history from the University of California at Irvine, where she graduated magna cum laude and Phi Beta Kappa, and she has a Ph.D. in social ecology with an emphasis on environmental analysis and design, also from the University of California at Irvine.

J. Frank Yates is an Arthur F. Thurnau professor, a professor of psychology, and a professor of marketing and business administration at the University of Michigan and a principal in the Psychology Department's Judgment and Decision Laboratory. He is also the coordinator of the Decision Consortium, which is a University of Michigan-wide association of faculty and students whose scholarship includes significant decision-making elements. The main focus of his research is on decision making, at both the theoretical and practical levels. That work has emphasized understanding how people decide in the challenging conditions of real life and developing means of assisting them to decide better in those circumstances. He is a past president of the Society for Judgment and Decision Making and is active in a variety of other efforts that are intended to advance decision scholarship, including efforts involving scholarly journals. He has been an active member of many government and other organizations, including the advisory panel of the National Science Foundation's Decision, Risk, and Management Science Program. He holds a Ph.D. from the University of Michigan.

Appendix B

Public Meeting Agendas

Meeting 1: Monday, November 22, 2010

9:45 am — Public Welcome and Study

- *Connie Citro*, Interim Deputy Director, Division of Behavioral and Social Sciences and Education
- *Clinton Oster*, Chair, Committee on the Effects of Commuting on Pilot Fatigue

Committee Member and Staff Introductions
Participant Introductions

10:00 am — Sponsor Perspective
Charge & Expectations of the Study

- *Dale E. Roberts*, Aviation Safety Inspector, Air Transportation Division, Federal Aviation Administration

Questions and Discussion

11:00 am — NTSB Comments

- *Mark Rosekind*, Member, National Transportation Safety Board

Questions and Discussion

12:00 pm — Working Lunch
Topics: Informal Discussion with Presenters

12:45 pm — Relevant Research
Flight Attendant Fatigue Study

- *Thomas Nesthus*, Civil Aerospace Medical Institute, FAA

Human Factors Monitoring Program: Fatigue Risk Management Scientific Study

- *Jessica Nowinski*, NASA Ames Research Center

(with Irving Statler participating via phone)

Questions and Discussion

2:15 pm — Stakeholder Comments

- *Charlotte O'Connell*, Pilot
- *Jeff Skiles*, US Airline Pilots Association

Questions and Discussion

3:00 pm End Open Session

Meeting 2: Monday, December 20, 2010

10:00 am Welcome and Introductions

- *Robert M. Hauser*, Interim Executive Director, DBASSE
- *Clint Oster*, Committee Chair

10:15 am Stakeholder and Public Comments: Part I

- *Captain (retired) Bill Mims*, A Pilot's Perspective
- *Steven Sargent*, Compass Airlines

11:15 am An Ongoing Study on Commuting and Pilot Fatigue

- *Lori Brown*, Faculty Specialist, Western Michigan University College of Aviation (via teleconference)

12:15 pm Working Lunch
Lunch will be served in the meeting room.
Topics: Discussion with Presenters

1:00 pm Stakeholder and Public Comments: Part II

- Airline Pilots Association, Intl., *Captain Bill Soer,* Flight and Duty Time Committee Member
- National Air Carrier Association, *George Paul*, Director of Technical Services
- Coalition of Airline Pilots Association, *Captain Bob Coffman*

2:30 pm General Discussion with Guests

3:00 pm Break

3:15 pm Learning Lessons for the Railroad Industry

- *Jeff Moller*, Assistant VP Operations Systems & Practices, Association of American Railroads, Washington, D.C.

4:15 pm Final Questions and Discussion

4:30 pm Adjourn Open Session